STUDENT
LAB NOTEBOOK

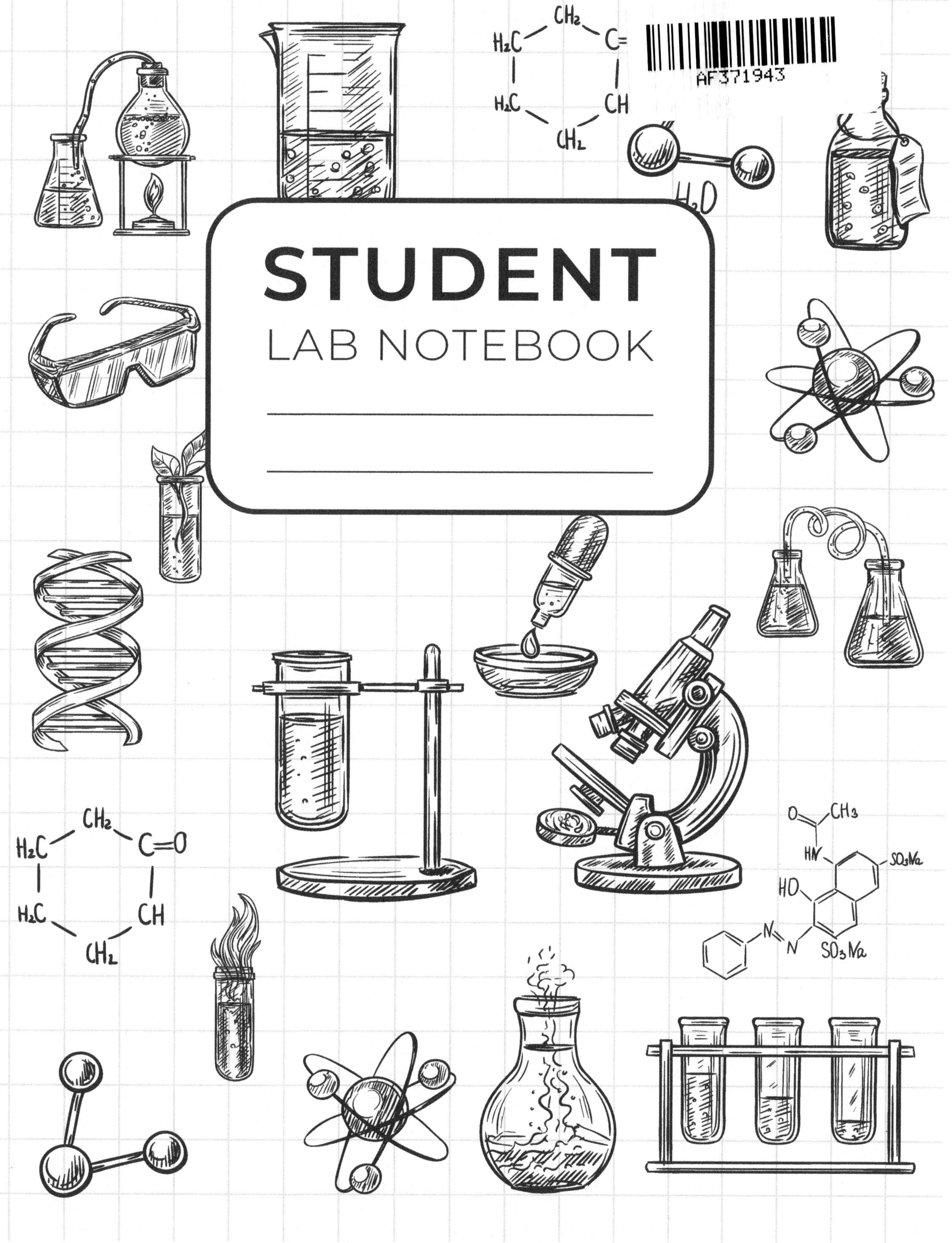

Table of Contents

No.	Date	Subject / Project / Experiment	Page

Table of Contents

No.	Date	Subject / Project / Experiment	Page
No.	Date	Subject / Project / Experiment	Page

Table of Contents

No.	Date	Subject / Project / Experiment	Page
No.	Date	Subject / Project / Experiment	Page

Table of Contents

No.	Date	Subject / Project / Experiment	Page

Title

Name

Date

Subject / Project / Experiment

Page

Title

Name

Date

Signature

Date

Signature

Date

<table>
<tr><td colspan="3">Subject / Project / Experiment</td><td>Page</td></tr>
<tr><td colspan="2">Title</td><td>Name</td><td>Date</td></tr>
</table>

<table>
<tr><td>Signature</td><td>Date</td><td>Signature</td><td>Date</td></tr>
</table>

Title

Name

Date

Subject / Project / Experiment

Page

Title

Name

Date

Signature

Date

Signature

Date

Title

Name

Date

Title

Name

Date

Signature

Date

Signature

Date

<table>
<tr><td colspan="2">Subject / Project / Experiment</td><td>Page</td></tr>
<tr><td>Title</td><td>Name</td><td>Date</td></tr>
</table>

<table>
<tr><td>Signature</td><td>Date</td><td>Signature</td><td>Date</td></tr>
</table>

Subject / Project / Experiment

Page

Title

Name

Date

Signature

Date

Signature

Date

Title

Name

Date

Subject / Project / Experiment

Page

Title

Name

Date

Signature

Date

Signature

Date

Title

Name

Date

Signature

Date

Signature

Date

Title

Name

Date

Signature

Date

Signature

Date

Title

Name

Date

<table>
<tr><td colspan="2">Subject / Project / Experiment</td><td>Page</td></tr>
<tr><td>Title</td><td>Name</td><td>Date</td></tr>
</table>

<table>
<tr><td>Signature</td><td>Date</td><td>Signature</td><td>Date</td></tr>
</table>

<table>
<tr><td>Subject / Project / Experiment</td><td>Page</td></tr>
<tr><td>Title</td><td>Name</td><td>Date</td></tr>
</table>

Signature   Date   Signature   Date

Title

Name

Date

Signature

Date

Signature

Date

<table>
<tr><td colspan="2">Subject / Project / Experiment</td><td>Page</td></tr>
<tr><td>Title</td><td>Name</td><td>Date</td></tr>
</table>

<table>
<tr><td>Signature</td><td>Date</td><td>Signature</td><td>Date</td></tr>
</table>

Subject / Project / Experiment
Page
Title
Name
Date
Signature
Date
Signature
Date

Title

Name

Date

Signature

Date

Signature

Date

Title

Name

Date

Title

Name

Date

Signature

Date

Signature

Date

Subject / Project / Experiment

Page

Title

Name

Date

Signature

Date

Signature

Date

Title

Name

Date

Subject / Project / Experiment
Page
Title
Name
Date
Signature
Date
Signature
Date

Title

Name

Date

Title

Name

Date

Signature

Date

Signature

Date

<table>
<tr><td>Subject / Project / Experiment</td><td>Page</td></tr>
<tr><td>Title</td><td>Name</td><td>Date</td></tr>
</table>

<table>
<tr><td>Signature</td><td>Date</td><td>Signature</td><td>Date</td></tr>
</table>

Title

Name

Date

Title

Name

Date

Title

Name

Date

Signature

Date

Signature

Date

Title

Name

Date

Signature

Date

Signature

Date

Subject / Project / Experiment

Page

Title

Name

Date

Signature

Date

Signature

Date

Title

Name

Date

Signature

Date

Signature

Date

Title

Name

Date

Made in the USA
Monee, IL
07 July 2026

56608341R00063